Civil Disturbances
St. Petersburg, Florida

SUCCESSFUL FIRE/EMS RESPONSE TO DISTURBANCES

October 24-25, 1996

November 13-14, 1996

Local Contact: Fire Chief James K. Callahan
St. Petersburg Fire and Rescue
400 9th Street South
St. Petersburg, FL 33701
(727) 893-7694

Department of Homeland Security
United States Fire Administration
National Fire Data Center

United States Fire Administration
Major Fire Investigation Program

The United States Fire Administration develops reports on selected major fires throughout the country. The fires usually involve multiple deaths or a large loss of property. But the primary criterion for deciding to write a report is whether it will result in significant "lessons learned." In some cases these lessons bring to light new knowledge about fire--the effect of building construction or contents, human behavior in fire, etc. In other cases, the lessons are not new, but are serious enough to highlight once again because of another fire tragedy. In some cases, special reports are developed to discuss events, drills, or new technologies or tactics that are of interest to the fire service.

The reports are sent to fire magazines and are distributed at National and Regional fire meetings. The reports are available on request from USFA. Announcements of their availability are published widely in fire journals and newsletters.

This body of work provides detailed information to the nature of the fire problem for policymakers who must decide on allocations of resources between fire and other pressing problems, and for the fire service to improve codes and code enforcement, training, public fire education, building technology, and other related areas.

The Fire Administration, which has no regulatory authority, sends an experienced fire investigator into a community after a major incident only after having conferred with the local fire authorities to ensure that USFA's assistance and presence would be supportive and would in no way interfere with any review of the incident they are themselves conducting. The intent is not to arrive during the event or even immediately after, but rather after the dust settles, so that a complete and objective review of all the important aspects of the incident can be made. Local authorities review USFA's report while it is in draft form. The USFA investigator or team is available to local authorities should they wish to request technical assistance for their own investigation.

For additional copies of this report, write to the United States Fire Administration, 16825 South Seton Avenue, Emmitsburg, Maryland 21727 or via the USFA WEB Page at http://www.usfa.dhs.gov/

U.S. Fire Administration

Mission Statement

As an entity of the Department of Homeland Security, the mission of the USFA is to reduce life and economic losses due to fire and related emergencies, through leadership, advocacy, coordination, and support. We serve the Nation independently, in coordination with other Federal agencies, and in partnership with fire protection and emergency service communities. With a commitment to excellence, we provide public education, training, technology, and data initiatives.

ACKNOWLEDGMENTS

The United States Fire Administration's Major Fires Investigation Team would like to thank the following people for their assistance with this report:

City of St. Petersburg

> **Mayor**
> David J. Fisher
>
> **St. Petersburg Fire and Rescue**
> James K. Callahan, Fire Chief
> James D. Large, Assistant Fire Chief/Chief Fire Marshal
> M. Thomas Burton, Assistant Fire Chief/Operations
> William Jolley, Chief of Rescue
> Douglas A. Lewis, Chief of Safety and Training
> Thomas T. Kitchen, District Chief
> Chris Bengtvengo, Public Education Officer
> The Officers, Firefighters, and Firefighter/Paramedics from Station 1
> The Officers, Firefighters, and Firefighter/Paramedics from Station 5
>
> **Emergency Management Office**
> Charles L. Whisenant, Administrative Assistant
> **Police Department**
> Darrel W. Stephens, Police Chief
> Goliath J. Davis Ill, Assistant Chief/Administrative Services
> Charles B. San Marco, Assistant Chief/Investigative Services

Lealman Fire Rescue

> Gary Wolff, Deputy Chief

Pinellas County 9-1-1

> Barry Mogai, Director, Emergency Communications/9-1-1
> Lori Collins, Director of Public Education

TABLE OF CONTENTS

ABSTRACT

This report is intended to provide information regarding the civil disturbances that took place in St. Petersburg, Florida, on the nights of October 24-25 and November 13-14, 1996, and detail how the city's fire and rescue services handled the crises. Most of the response duties in St. Petersburg related to civil disturbance duties were centered on law enforcement agencies. However, the St. Petersburg Fire and Rescue Department (SPFR) played a significant role in the outcome of the two 1996 disturbances. The community suffered significant fire loss, but there were very few civilian injuries and only four minor firefighter injuries over the course of the respective incidents.

ABSTRACT

OVERVIEW

The City of St. Petersburg is located in the middle of Florida's Gulf Coast. The city covers approximately 60 square miles and has an estimated population of 256,000. The St. Petersburg Fire and Rescue Department has approximately 370 firefighters and paramedics serving from 13 stations. The department provides fire, rescue, and emergency medical services with 13 engine companies, four truck companies, and 10 EMS units. It is a career municipal department with a budget of $22 million budget.

On October 24, 1996, and on November 17, 1996, the city experienced two successive civil disturbances. The first one was triggered by an incident involving a Caucasian police officer who fatally shot an 18-year-old African-American man during a traffic stop. The second disturbance occurred after the police officer was cleared in the fatal shooting. The city estimated that over 60 separate arson fires were set during the two disturbances, causing approximately $6 million in property damage and untold economic losses for businesses that were forced to close. Most of the response duties related to the civil disturbances were handled by law enforcement agencies. The St. Petersburg Fire and Rescue Department (SPFR) played a significant role in the outcome as well, since dozens of criminal fires were set that resulted in significant fire loss.

The St. Petersburg's Fire and Police Departments had an operational pre-incident plan for dealing with civil disturbances. The plan proved effective in helping to mitigate the consequences stemming from the outbreaks of violence, looting, and firesetting. For example, independent fire units were formed into task forces consisting of two engine companies and a district chief, supplemented with truck company personnel. The truck companies were left in the staging areas. The task forces responded to and handled the incidents as a single operational entity. In a significant departure from the established protocols, however, the fire task forces operated at a distance from police activity whenever possible, and responded to the incidents separately from the police. Only after the police had secured an area were the task force units allowed to advance. In addition, the EMS units received police escorts throughout the disturbances.

To deter angry mobs from directing their hostility towards fire personnel as well, police and fire units were deployed separately. This contrasts with the experience in Los Angeles during the disturbances there, in which three firefighters were shot and all fire task forces had to be protected by police and the National Guard.

This report reviews St. Petersburg's pre-emergency plan for civil disturbances. It provides an overview of the factors that led to the first disturbance, the response, and the changes and preparations that were made immediately thereafter and during the second disturbance, three weeks later. The report also highlights the lessons learned from these two events:

- Pre-incident planning was essential to the successful response;

- Close fire and police communications was essential to ongoing fire department operations; and

- The positive community image of the fire department helped make it possible for firefighters and paramedics to operate without becoming targets of violence.

The St. Petersburg response mirrored in many ways the recommendations of a 1994 United States Fire Administration Report (FA-142) on responding to civil disturbances. That report offers recommendations for organization and operations during a civil disturbance, based on incidents of civil unrest during 1992 in the United States.

Early in the first incident, SPFR developed a system to prioritize responses. The first priority was safety for all responders, with no exceptions. Four main topics were addressed:

1. Unified command was implemented immediately, which was an essential step in coordinating tactical activities. Early in the situation, the Operations Chief initially requested a liaison from the St. Petersburg Police Department, but a County Sheriffs Department sergeant was assigned instead. The Sheriff had access to more resources throughout the entire County and thus was able to augment the City's resources for response.

2. Frequent briefings and planning meetings took place between senior police and fire officers during the civil unrest. Fire and Police leadership and field commanders were informed continuously of the progress of tactical strategies and of the status of firefighters and police officers in the field. Situations in civil disturbances can change quickly and may take a turn for the better or the worse—depending on the environment and the mood of civilians. Accordingly, tactical measures were changed and implemented on short notice.

3. Communications management was a crucial component of dispersed tactical activities. Dissemination of the wrong information by the media or by the Public Information Officer (PIO) could have incited destructive activity by troublemakers and made emergency situations less manageable. The media in these situations can support ongoing operations by providing timely warnings and helping to dispel rumors; or it can contribute to misinformation or speculation. Accurate, timely, and consistent information from designated public information officers helps the media present an accurate picture to the public. Emergency communications were very well handled in these incidents, according to the Operations Chief, who said, "Having common equipment, radio frequencies and terminology, and daily experience with automatic aid was, without question, a blessing." (Pinellas County consists of 21 fire and rescue organizations, all of which adhere to the automatic aid system.)

4. Mutual aid agreements were activated, and all participating organizations were called to respond on short notice and to report to the staging area to support the fire and police task forces. The Incident Commander (IC) requested the services of EMS personnel within the SPFR, private county EMS, politicians, local, county, Regional, and state resources, public works departments, community leaders, and hospitals.

These four factors formed the basis of an efficient response system, and were critical to fire and rescue operations during the first incident.

SUMMARY OF KEY ISSUES

Issues	Comments
Comprehensive Emergency Management Plans	The City of St. Petersburg, and the fire and the police departments had plans and procedures in place to manage and control civil disturbance, which proved effective.
Mutual Aid	Through established Pinellas County resources and automatic mutual aid agreements, fire and police received timely back up and support with staffing and equipment.
Incident Management	Use of incident command practices allowed for smooth operations between the fire and police units, as well as an organized flow of information through the Emergency Operations Center. There were regular updates on the situation status for all the city agencies and the news media.
Protection of Emergency Responders	Fire forces did not respond into hostile locations immediately but waited until the police first secured the areas. Police escorts accompanied Advanced Life Support (ALS) units.
Firefighting Strategy and Tactics	The normal firefighting response was adjusted to a task force model. A group leader was placed in charge of each task force. Task forces used quick attack, hit-and-run tactics to knock down fires rapidly and protect exposures. They moved quickly from incident to incident, postponing overhaul operations.
Separation of Fire and Police Units	The separation of fire units from police activity minimized fire and EMS responders from becoming the targets of crowd violence during the disturbances. The positive image of the fire department that had existed in the minority community also contributed to the fire department's safety.

HISTORICAL PERSPECTIVES AND PRE-INCIDENT PLANNING

The City of St. Petersburg is the largest of 24 municipalities in Pinellas County, and is located in approximately the middle of the Gulf Coast of Florida, near Tampa. Light industry, military, tourism, and a large retirement community dominate the economy. Approximately 17 percent of the residents are minorities, most of whom occupy the historically African-American southern section of the city. This area of St. Petersburg has experienced many of the same problems that afflict other minority communities across the country, including unemployment, crime, drugs, and poverty.

Prior to the disturbances, both the police and fire departments were operating prevention and outreach programs in the south section of St. Petersburg. The police department had ongoing community policing and a neighborhood watch program in place. The fire department offered several fire prevention and education programs through churches, schools, libraries, and recreation centers.

With the one exception of the activities of the National People's Democratic Uhuru Movement, a small black advocacy group based in St. Petersburg, the city's race relations were calm. The political landscape included minority representation and was non-confrontational. The last time St. Petersburg had experienced civil unrest was in 1968 after a garbage strike.

On October 24, a Caucasian police officer shot an 18-year-old African-American man during a traffic stop--a short distance from the National People's Democratic Uhuru Movement's, an African American separatist group's headquarters. Arriving on the scene quickly, some group members took advantage of the situation to voice their anger over what they perceived to be racist treatment of minorities by the police. City officials believed that the early comments by National People's Democratic Uhuru Movement members, along with media portrayals of the shooting, led to a quick escalation of hostilities in the first disturbance. Most of those arrested were youths around the age of 15.

Planning--The City of St. Petersburg has a comprehensive *Disaster Operations Plan*. The Plan is reviewed annually and is updated every two years. The Plan describes the key operational elements for emergency response, such as county and city law enforcement, fire service coordination, dispatch, communications, and multi-jurisdictional response. Though frequently used for weather emergencies, the "Plan" had not been field-exercised for civil unrest.

According to the St. Petersburg Plan, overall emergency management and coordination was to be conducted from the city's Emergency Operations Center (EOC). The Plan called for the activation of the Center upon declaration of an emergency, or during an incident of sufficient scope where the support of the EOC is appropriate. Representatives of the city's agencies--including the mayor, fire chief, and police chief--respond to the EOC to assist in incident management. The EOC was activated during both disturbances.

The St. Petersburg Plan also directed that the Incident Command System (ICS) be used to manage and coordinate major incidents. The system focused on four key elements: development of incident strategy and tactics; resource management; interagency and interjurisdictional coordination; and communications. The plans and procedures, although untested at the time, contained the essential elements to handle the civil unrest that challenged St. Petersburg's emergency services.

St. Petersburg also has a *Tactical Plan for Civil Disturbance*, developed in 1994, which designated the Police Department as the lead agency, with the fire department providing suppression and EMS. Florida law requires that law enforcement agencies assume the lead role in the event of a civil disturbance.

In St. Petersburg, the fire chief also serves as the Director of Emergency Management and is assisted by two full-time employees assigned to the office, which is responsible for developing disaster plans and managing the Emergency Operations Center located at SPFR headquarters. Most of the emergency planning and preparedness in St. Petersburg is focused toward the annual hurricane season. The Office of Emergency Preparedness does not function as a separate full-time office per se but, rather, is housed in and operates through the fire department. It works closely with the Pinellas County Emergency Management Office, a fully staffed office located in the county seat of Clearwater.

The Pinellas County Sheriff's Department and other fire and EMS agencies supported the St. Petersburg Fire and Police Departments during the civil disturbances. The 21 fire departments and 17 rescue/ EMS agencies throughout Pinellas County routinely coordinate emergency operations through an automatic mutual aid program. The SPFR is a member of the County Operation Group that meets and trains regularly. The Group's experience with mutual aid and integrating operations enabled rapid augmentation of resources when SPER responded to the incidents. The Fire Operations Officer at the tactical command post commented, "The County is functionally consolidated with regard to command, standard operating procedures, communications, and training."

Fire Department Role--As outlined in the earlier USFA report, most jurisdictions assign police escorts to responding firefighting units if there is civil unrest--an approach that essentially merges the two entities into one public safety unit. In contrast, during St. Petersburg's two episodes of civil unrest, fire department operations actually operated separately from police whenever possible. The City's plan called for police to clear all fire locations and establish a safe haven for fire operations prior to dispatch. Had an area become unstable, fire companies were to be diverted out of the area. The protocol was used to maximize firefighter safety during tactical operations. Police escorts, however, were used for EMS calls to protect paramedic crews and their patients.

The SPFR activated its callback policy to bring in off-duty personnel. Personnel were requested to report to their stations and the staging areas; other SPFR members reported on their own to their assigned stations and staging area after hearing of the disturbance on the radio or television. Additional ALS staff for the higher demands were obtained from automatic mutual aid within the county and outside the county, including the Tampa Fire Department.

The fire department's response to the disturbances embraced the following key elements, essential to the successful management and mitigation of any emergency incident:

Risk Analysis: Fire department field personnel had the training and competency to identify the potential for escalating violence and quickly notified top department officials of the situation through the chain of command.

Operational Planning: Command officers immediately identified the need to adjust standard operations to handle the disturbances. Operational needs were evaluated and plans to handle the emergency were put into action. Operations involving civil unrest incidents were assigned differently from the normal city response to fire and EMS calls.

Command Structure: The Incident Command System was immediately implemented. Using existing city and fire department plans, appropriate roles and responsibilities for all departments and agencies were assigned.

Tactics: Working through the City's Emergency Operations Center (EOC), Police, Fire and EMS coverage was maintained throughout the city. Response procedures and fireground tactics were adjusted to address safety issues.

In each incident, safety of fire and EMS personnel was a high priority. In reviewing the Fire Department's *Civil Disturbance Operations Procedure Guidelines* (see Appendix D), the safety of personnel; coordination with police, command structure, and communications; and fireground tactics were all addressed. The plan incorporated the guidelines and principals of National Fire Protection Association (NFPA) 1500, Section 6.7, for fire department response to civil disturbances.

THE DISTURBANCE OF OCTOBER 24

The first major civil disturbance spread across the city's south side shortly after 5:30 p.m. on October 24, 1996. As noted earlier, the disturbance ensued after a Caucasian police officer shot an 18-year-old African-American man who had been stopped for a traffic violation. Within 30 minutes of the shooting, over 100 angry protesters gathered at the scene, including members of the National People's Democratic Uhuru Movement. Led by a few young instigators in the crowd, community members became increasingly violent. Scuffles started with police officers, including the police chief who had responded to the scene. A barrage of rocks, bottles, stones, Molotov cocktails, and anything else that could be thrown was directed toward the police. Events accelerated so quickly that notification of the imminent potential for civil unrest had not yet been received by other city officials, nor by the police or Fire Communications Center.

Several vehicles, including a news van, were set on fire, precipitating the fire department's involvement. St. Petersburg Engine 5 was dispatched to the initial fire. Engine 3 and a District Chief were in the area and responded to the scene after hearing the dispatch. When Engine 3 arrived on the scene, a riotous crowd broke the windows of the fire engine while the crew scrambled for safety. Both Engine 3 and Engine 5 retreated and reported the situation to their Communications Center. The District Chief was forced to abandon his command vehicle in the midst of the crowd.

In accordance with the fire department's situation reports, the police officers on the scene passed along dated reports on the growing magnitude of the incident. The Mayor of St. Petersburg was advised of the situation and a state of emergency was declared at 2108 hours, activating the City's Disaster Operation Plan. At the same time, the EOC, located at fire department headquarters, was fully activated and staffed.

Under the Fire Department guidelines for civil disturbances, *Civil Disturbance Operations Procedure Guidelines*, commanders adjusted fire and EMS response procedures to maintain service delivery while increasing safety for personnel. Fire and EMS resources were organized into Task Forces and initiated a dispatch system from a tactical command post located outside the perimeter of the affected area. All EMS units increased staffing from two to three paramedics. The increased staffing of EMS units came from move-up companies, from the now out-of-service truck companies, and from off-duty personnel.

Consistent with the *Tactical Plan for Civil Disturbances*, the St. Petersburg Police Department became the lead agency for controlling the disturbance. They cleared all fire areas of hostile activity prior to dispatching task forces. Police escorted EMS units from time of dispatch until individual runs were completed.

The state of emergency declaration ended at 0345 hours on October 25. City police and fire personnel, as well as the Pinellas County Sheriffs Department, remained on alert for several days with additional staffing and resources. The Florida National Guard continued to stand by at the Tactical Command Post located in the Thunder Dome[1] through Sunday, October 27.

Personnel Safety

At the onset of the unrest on October 24, the Command Officer, after attempting to extinguish a number of the fires, relocated to a safe site at Thunder Dome. Shortly after this retreat, a plan was launched to identify the areas of instability (called "hot zones") and to alter the responses and activities in such areas. No further fire and rescue personnel were sent in until such time as law enforcement could confirm that the scene was secure.

While many of the Advanced Life Support (ALS) units had bullet resistant vests for personnel, none of the fire suppression units had this protection. Vests from ALS units that were not involved in the hot zone were secured to ensure that all rescue team members had vests. Except for the vests, the

---

[1]The Thunder Dome has since been renovated and renamed Tropicana Field. For purposes of this report, it will be referred to as the Thunder Dome.

firefighters had no equipment designed to protect them from violence. Standard firefighter gear provides only minimal protection from hurled rocks and bottles, and no protection from gunfire.

Upon arrival at the staging area, crews were assembled into task forces. These generally consisted of two engine companies, one support unit, and one Chief Officer who served as a task force leader. Task forces were to travel as a team, work as a team, and return as a team. Whenever possible, fully enclosed cabs were used to provide safer travel.

After being assigned to a task force; crews were briefed on the conditions they might encounter and given both tactical and safety instructions. As crews were assembled into task forces, task force leaders were given the ultimate decisionmaking authority to retreat from any call that they determined to be unsafe or unstable. The concept of accountability and the need to look after one another was stressed.

Throughout all of the activity, SPFR personnel were successful in limiting injuries. Three St. Petersburg firefighters and one Lealman firefighter sustained minor injuries. Table 1, Summary of Injuries to Personnel, shows the injuries received by response personnel.

Table 1. Summary of Injuries to Personnel

Date	Location	Description
10/24/96	16th St. and 18th Ave. S.	A firefighter was working at a rescue scene when pieces of concrete block were thrown at the rescue crew. Although the crew tried to avoid being hit, a firefighter was struck in the head. He received treatment at Edward White Hospital and was released.
11/13/96	Queensboro Ave. between 43rd and 44th Sts. S.	A lieutenant was working at the scene of a fire involving three (3) houses; high winds blew embers from the involved houses, which hit the lieutenant on the right side of his face. He was working outside of the structure at the time of the injury. He did not need treatment for injury.
11/14/96	4300 Queensboro Ave. S.	A firefighter was assigned to ride with District 18 in Task Force 3. While working at a fire, the firefighter reported that he inhaled some smoke. There were not enough air packs on scene for all personnel. He did not need treatment for his injury.

In addition, a firefighter from Lealman Fire and Rescue required treatment at the scene for debris in one eye. Map 2, Map of St. Petersburg Streets, shows where in St. Petersburg the riots generally occurred; this map also illustrates the locations of the command post and firefighter injuries.

Over the course of the disturbances in late October and mid-November, 78 arson and attempted arson fires were set in commercial structures, both occupied and unoccupied dwellings, schools, vehicles, and government offices. Map 3 depicts the area where these fires were concentrated and a list of the targets.

Map 1. State of Florida

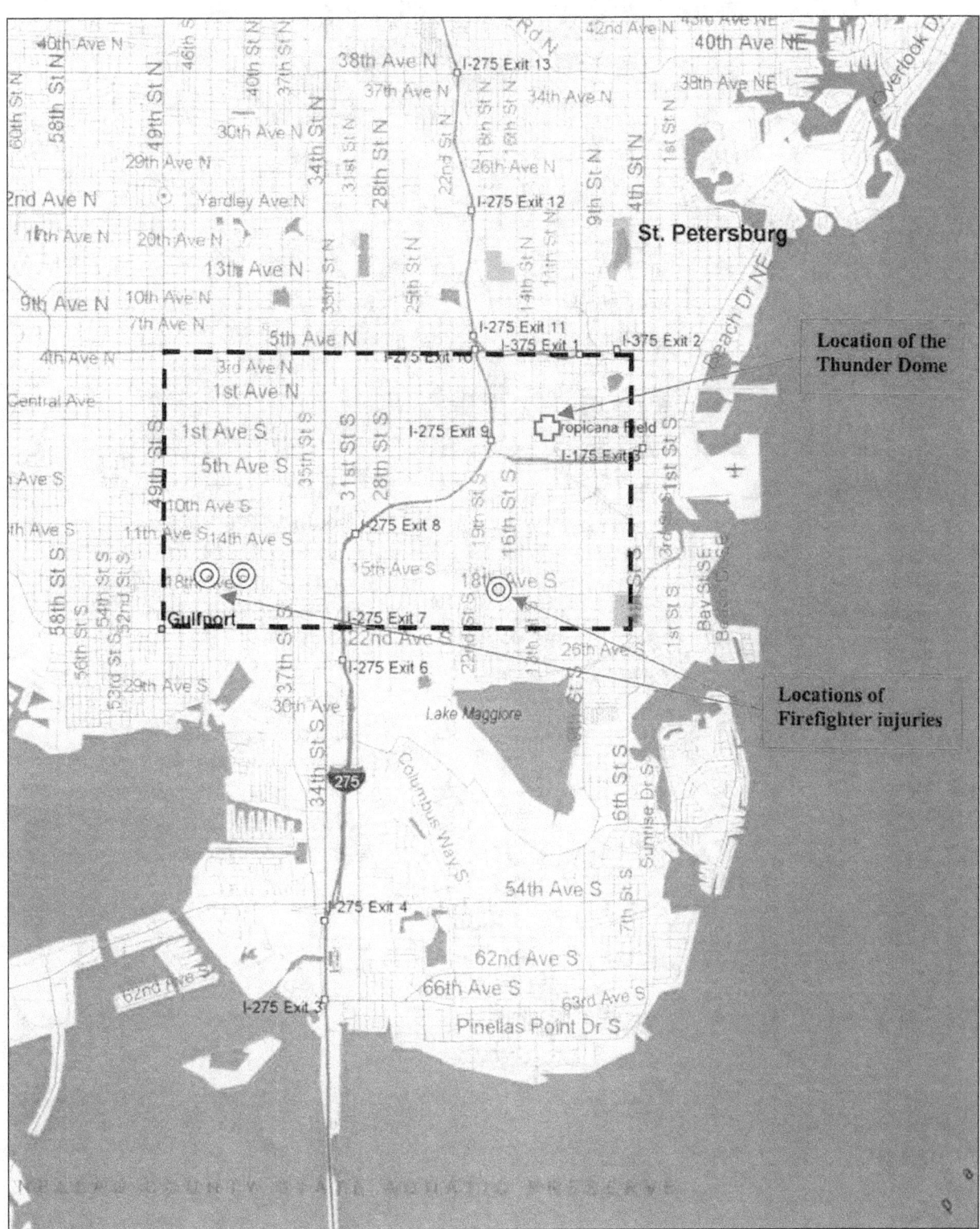

Map 2. Streets of St Petersburg FL

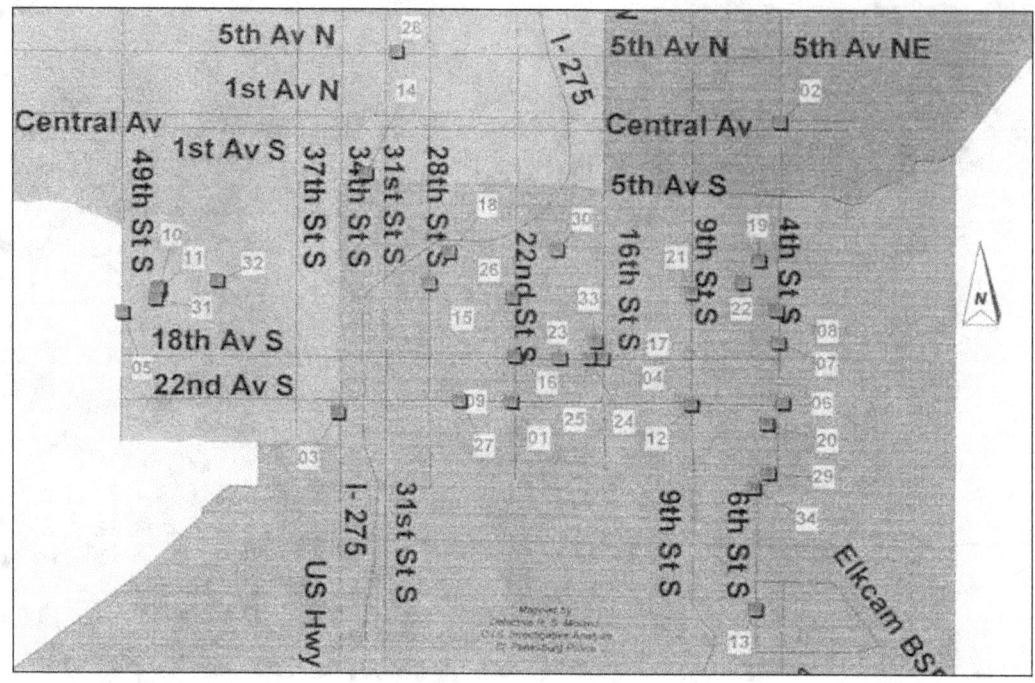

NUM	Offnmbr	Name	Address
01	96-059265	Carter's Florist	2200 22nd Av S
02	96-059168	U.S. Post Office	76 4th St N
03	96-059270	Kidney Foundation Thrift Store	2333 34th St S
04	96-059299	Vehicle fire WTOG TV Inc	18th Ave S && Preston St
05	96-060004	Georgia Meat Market	1500 49th St S
06	96-059277	Chattaway's Restaurant	358 22nd Ave S
07	96-059279	Mast's Bait and Tackle	1700 4th St S
08	96-060005	Police Resource Center	1453 4th St S
09	96-059294	Probation & Parole	2554 22nd Ave S
10	96-059241	Unoccupied dwelling (CWA CASE)	4639 13th Av S
11	96-059242	Unoccupied Dwelling	4646 13th Ave S
12	96-059892	Badcock Furniture Co.	2200 9th St S
13	96-059205	Suntrust Bank	4142 6th St S
14	96-059310	Unoccupied Dwelling	3218 4th Av S
15	96-060006	Wildwood Meat Market	1228 28th St S
16	96-059280	Sun Liquors	2205 18th Av S
17	96-060007	Vehicle fire, Police cruiser	18th Av S && 16th St S
18	96-059201	Unoccupied Structure	955 26th St S
19	96-060009	Doctor's Office	500 10th Av S
20	96-060010	Trinity Church (CWA CASE)	2401 5th St S
21	96-059288	Accurate Locksmith	1326 9th St S
22	96-060012	Unoccupied Duplex (638-640)	638 12th Ave S
23	96-060013	18th Avenue Supermarket	1856 18th Av S
24	96-059299	Vehicle fire, WTOG TV Inc.	18th Ave S && Preston St
25	96-059301	Vehicle Fire	18th Ave S && Preston St
26	96-059318	Crossroads II Store	1317 22nd St S
27	96-059393	Probation and Parole	2554 22nd Av S
28	96-059399	Glidden Paint	3001 5th Av N
29	96-059394	Vehicle Fire	3180 5th St S
30	96-059275	Three Brother's Grocery (Attempt)	1906 9th Av S
31	96-059388	Unoccupied Dwelling (Attempt)	4663 14th Ave S
32	96-059609	Shed	4260 12th Ave S
33	96-059489	Occupied Dwelling(CWA Attempt)	1722 Prescott St S
34	96-060074	Don's Irish Pub (Attempt)	3007 6th St S

Map 3. This map depicts south St. Petersburg, Florida. The offset dark squares are the arson locations during the period of October 24–27, 1996. The numbers on the map relate to the number on the corresponding list, which identifies the type of property. The numbers do not reflect the order of arson occurrences.[2]

[2] Map and incident information courtesy of Detective R. S. Moland, C.I.S. Investigative Analysis, St. Petersburg Police department.

Command Post Operations, Communications, and Dispatch--The police and fire departments each established a tactical command post in the parking lot of the Thunder Dome, a multi-purpose sports arena with acres of parking. The complex provided a safe haven for emergency personnel north of the disturbance. The parking lot provided excellent access to the emergency scene and sufficient space for staging equipment and personnel.

Communications were maintained at the tactical command post between the EOC, operating units, and the department's logistical support teams. Communications with fire units were maintained through an 800 MHz trunked radio system, which is capable of isolating emergencies on separate channels as conditions dictate. Several radio channels were dedicated to handling the disturbance, including a citywide emergency band for police and fire operations that was used during the October 24 incident.

Police and fire emergencies are normally dispatched through separate systems within the City of St. Petersburg. During the disturbances, the Countywide system of dispatch and automatic mutual aid addressed the ongoing need for Police, Fire and EMS services. Move-ups and transfers of companies are a routine part of the day-to-day operations of the system, and several companies were transferred to fill in stations during the disturbances.

As the incident escalated, police switched to a three-channel special event frequency that was not available to the St. Petersburg Fire and Rescue units. The change in police communications temporarily isolated fire personnel, resulting in a delay in dispatching fire equipment--the only notable breakdown in the overall plans and operations during the first disturbance. The communications problem was immediately identified by both Fire and police officials. Prompt dispatching of fire units resumed after a police liaison arrived with a portable police radio at the Fire Tactical Command Post.

Other Agencies--All responding outside agencies were directed to report to the appropriate tactical command post. A helicopter from the Pinellas County Sheriffs Department was available and used during the first incident for aerial surveillance. The St. Petersburg plan provided for the support of the Florida National Guard. On the evening of the October 24 disturbance, the Florida Governor offered State support to local officials. The Florida National Guard unit from Tampa was activated, and Guard units were stationed at the Thunder Dome after the first disturbance, arriving early Saturday morning, October 25.

Firefighting Response and Tactics--The City of St. Petersburg has two fire divisions, North and South, each commanded by a District Chief. Operating throughout the city are 13 engines, 4 ladders, a rescue company, and 10 transport ALS units. The county, however, is supported by a private EMS system. County policy is that after 10 minutes have elapsed, the City's Fire ALS unit may transport.

As noted earlier, SPFR organized fire operations into a task force approach, combining individual companies into larger task forces that would respond as a single entity to handle large incidents. Each task force included a command officer as the designated task force leader and at least two engine companies with eight or nine personnel assigned. In addition, all the ladder companies in the disturbance area were deactivated, and personnel were reassigned to increase staffing for the task force and EMS response. The SOPs were modified according to the intensity of the disturbance. For example:

- No firefighters working on roofs

- No laddering of buildings

- No ladder truck companies in impact zones

- No overhauling of fires

- No interior fire attacks

- No wearing of SCBAs

The main reason for avoiding the use of ground ladders and towers was to prevent firefighters from becoming vulnerable, easy-to-see targets. Also, truck companies would find it difficult to suspend operations fast enough for a rapid withdrawal if danger threatened.

Firefighting tactics also changed to address the situation. Any fires that endangered occupied dwellings were given the highest priority. A 'hit and run' approach was adopted, using mounted master streams with smoothbore nozzles from tank water. All the engines had fixed deck guns, which became the tool of choice for rapidly attacking the fires. All equipment on the exterior of the engines was removed to prevent it from being stolen or used against fire crews. All responses into the disturbance area were made without lights and sirens. Supply lines were only laid after the task force leader determined that exposures were in fact threatened and that conditions were stable. No laddering of buildings or interior firefighting was allowed in the area of the disturbance, to reduce the exposure of firefighters and to allow for rapid withdrawal.

All task forces were equipped with common channel portable radios. Additionally, at least one unit from the St. Petersburg Fire and Rescue was part of each task force, to ensure knowledge of the area for response as well as retreat, if necessary.

Prior to dispatch, each task force was briefed at the staging area. All personnel were informed of the circumstances and the environment they would be encountering. They were informed that the St. Petersburg Police Department would check each response area before fire task forces would be dispatched. Task force leaders were directed to assess each fire location on arrival for the safety of fire personnel and to monitor all safety issues throughout the entire operation. **If the task force leaders believed that crews were in danger, they had the authority and responsibility to disengage from operations and leave the area.**

Each of the task forces was to function as a single unit. They were to operate exclusively on the assignment to which they were dispatched. Upon completion of each assignment, they were to leave the area immediately and return to the Fire Tactical Command Post, reporting their availability to dispatch. Upon returning to the staging area, the task forces would be debriefed relative to their assignment, area conditions, injuries, and fitness to return to duty.

EMS Response--During normal EMS operations in the incorporated areas of the county, a fire EMS unit responds with a faster response time. The Fire EMS units stabilize, pack, and maintain control of both the scene and the patient until a county private EMS provider arrives for transport. If the private ambulance is delayed and waiting time equals or exceeds ten minutes, the city's EMS unit is authorized to transport during normal operations.

Staffing on SPFR advance life support (ALS) units was increased from two to three paramedics. Owing to the small number of personnel assigned to each medic crew, as well as the nature of EMS operations, Police escorted each medic unit. Police remained on location while patients were assessed and prepared for transport, and then escorted the unit out of the area to the triage area located at the Thunder Dome. These units--SPFR personnel, working with the police--performed "Hot Zone" EMS under the direction of the fire department and provided transport at the triage area. The county's private EMS providers (which were not permitted to operate in the emergency areas) then transported casualties to area hospitals. SPFR maintains a day-to-day working relationship with the county's private EMS system, so this transition from hot zone, triage area, and the hospitals went well.

The SPFR medical group established three extraction teams to respond to calls generated by the incident as well as to 'regular' calls in the affected area. The extraction teams consisted of two SPFR paramedics and at least one SWAT medic from either the St. Petersburg or Pinellas County SWAT Teams. The EMS personnel were escorted to the call locations by at least one patrol unit staffed with four officers. The patients were transported from the riot area in fire department vehicles to a transport staging area where the patient was loaded off from the rescue vehicle and loaded into the ambulance transport vehicle. The cooperation between fire department personnel and transport provider personnel facilitated timely access to medical care.

The following night, EMS response was fashioned in the same manner as the previous night; however, six units were prepared to respond. The night was relatively uneventful, and only a small number of calls were generated from the special response area.

BETWEEN THE DISTURBANCES: FINE TUNING THE OPERATIONS

The existing Emergency Operations Plans for Police and Fire response to civil disturbances worked extremely well in St. Petersburg. The activation of the Emergency Operations Center, the incorporation of all resources of local government, and the use of the forward Tactical Command Posts worked to near perfection. The St. Petersburg response was an excellent example of an overall, integrated emergency management system.

In the short time period between the first and second disturbances, the St. Petersburg authorities had little time to conduct a detailed review of operations. However, prior to the second riot, which occurred three weeks later, the city was able to fine-tune some of the operational issues that occurred during the first disturbance. The lessons learned were gleaned from the first incident and from a documentation officer who was appointed early into the start of November's civil unrest. Lack of documentation during the first two nights of unrest had left SPFR with the enormous task of trying to sort out which units had been assigned to which incident and what resources had been used. Nine task forces were running at the peak of operation, all requiring documentation. Debriefings were recorded on individual sheets and labeled with the task force number and incident address. Incident numbers were assigned later. Besides enforcing an accountability system at the task force level and helping to avoid duplicate dispatching on calls, the documentation served longer-term benefits. It was used to apply for reimbursement by FEMA under the Stafford Act, and to assist business owners with the Small Business Administration loan process.

Tensions in the City were raised as the victim of the shooting was buried on November 2nd. The funeral procession was routed through the impact area, but there were no reports of civil unrest following the funeral. As the City Council wrestled with the aftermath of the first incident, Federal officials began a review of the events, including the minority community's reaction, Police response, and ongoing relations between the city establishment and minority residents. Economic and social issues were debated in various arenas.

An underlying theme from all public safety individuals interviewed during this investigation was that the media's coverage tended to distort the actual facts of the incident and contributed to the escalation of the disturbance. Most officials agreed that the media played a substantial role in keeping the shooting, civil unrest, and racial tension a focal point of news. Reporting on the disturbance of October 24 started with an initial flash report, with further reporting that played on the emotions and tension of the erupting crowd.

A key part of the city plan was recovery from the civil unrest. The Public Works Department initiated immediate cleaning of all streets within the affected area. Six tons of debris were removed in the five-square-mile area of the incident. The city ordered that abandoned or burned vehicles be removed. By sunrise the morning after each incident, the physical evidence of the disturbance, aside from the fire-damaged structures, was gone. The public works crews cleaned up debris that could be used to attack police and fire personnel who feared being ambushed. Rioters stashed arsenals of rocks, bottles, bricks, and stones at specific locations within the area of the first disturbance. The potential "weapons" were removed three or four times, only for the caches to be quickly restocked by troublemakers.

The city prepared itself for the November 14 release of the grand jury inquiry into charges against the police officer involved in the shooting. Also, the police department's internal report on the incident was scheduled to be released at the same time as the grand jury findings.

During the period between the first and second incident, police maintained surveillance of the individuals who were believed to have had instigated the first civil unrest. Planning occurred on a daily basis for many city and county agencies. Briefings and intelligence were shared by all affected agencies.

The night of the wake and funeral for the victim of the shooting, Halloween, and Mischief Night, the evening before Halloween, were times of heightened concern; however, no incidents were reported. Such was not the case when the grand jury and the St. Petersburg Police Department released their findings. Police intelligence reports indicated that the National People's Democratic Uhuru Movement would protest with possible violence when the anticipated results of each finding were made public. By combining the release of the grand jury inquiry with the St. Petersburg Police Department's report, the city hoped to condense the reaction into one event.

THE DISTURBANCE OF NOVEMBER 14

The second civil disturbance, November 14, 1996, began after the grand jury released its findings, which determined that no criminal charges should be filed against the Police officer involved with the shooting. The St. Petersburg Police Department found that several safety procedures had been violated and that the officer would be suspended without pay for 60 days.

Prior to the grand jury's announcement, and in an effort to preempt violence, the St. Petersburg Police Department attempted to arrest members of the National Democratic People's Uhuru Movement for outstanding warrants. Police were confronted by an angry mob, and a second episode of violent confrontations ensued.

The second disturbance was smaller and was handled with city resources because of the experience three weeks earlier. The city's emergency services were well prepared. A declaration of emergency was not deemed necessary. The second disturbance mirrored the first in that it was violent, but it was localized, isolated, and addressed quickly. As a result, the disturbance ran out of momentum in short order. Map 4 shows the location of arson incidents during this disturbance.

One key difference in the second incident was that rioters used automatic weapons stockpiled after the first disturbance, so the Pinellas County Sheriffs Department helicopter could not be used for aerial support and reconnaissance. The firing of automatic weapons could be heard throughout the night. Commenting on the differences between the two disturbances, a St. Petersburg Police Department Officer stated that the second disturbance seemed more like urban gorilla warfare.

During the November 14 disturbance, the Florida National Guard was placed on alert but remained at its base in Tampa, approximately 20 miles northeast of St. Petersburg.

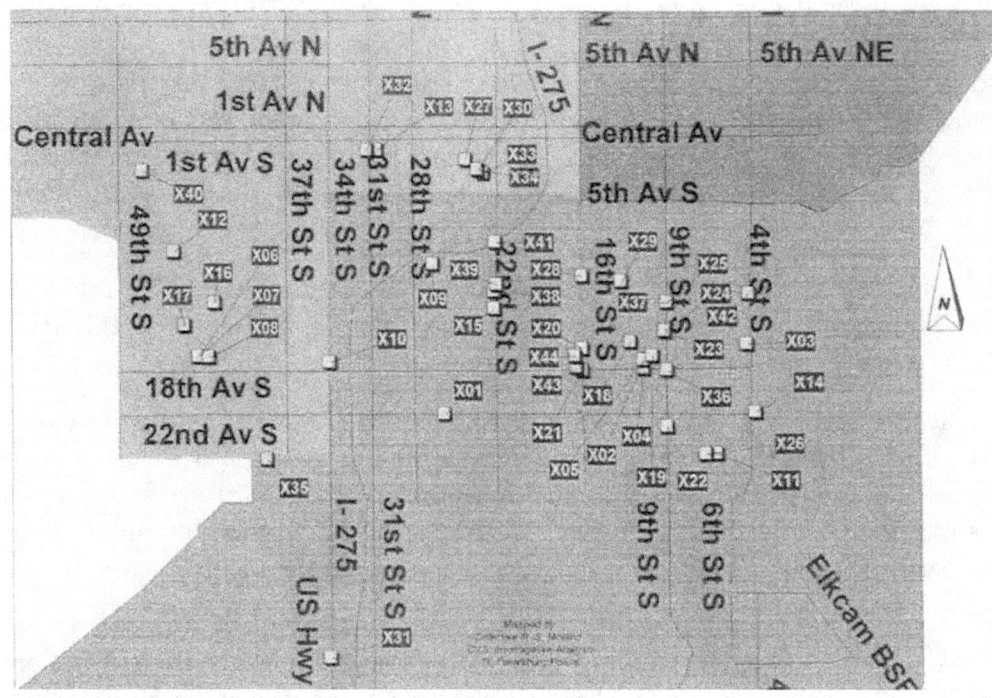

NUM	Offnmbr	Name	Address
X01	96-063076	Probation and Parole	2554 22nd Ave S
X02	96-063009	Unoccupied Dwelling	1721 Jewell St S
X03	96-063001	Vacant Commerical Building	426 Preston Ave S
X04	96-063875	Garage (O#96-63009)	1035 18th Ave S
X05	96-063009	Unoccupied Dwelling	1729 Jewell St S
X06	96-063881	Unoccupied Dwelling	4327 Queensboro Ave
X07	96-063881	Unoccupied Dwelling	4325 Queensboro Ave
X08	96-063881	Unoccupied Dwelling (1/2 Add)	1735 43 St S
X09	96-062942	Department of Juvenile Justice	955 26 St S
X10	96-063876	Chevron Gas Station	1750 34th St S
X11	96-062917	Vehicle (O#96-63021)	675 26th Ave S
X12	96-063023	Action Auto Sales	4545 8th Ave S
X13	96-062969	Puckett's retail store	3050 1st Ave S
X14	96-062986	Vehicle	416 22nd Ave S
X15	96-062990	Crossroads grocery store	1317 22nd St S
X16	96-063877	Unoccupied Dwelling (O#96-62930	4263 13th Ave S
X17	96-063002	Occupied Dwelling (One Burn Victim)	4425 15th Ave S
X18	96-063003	Dwelling	1066 16th Ave S
X19	96-062998	Southside Fundamental School	1701 10th St S
X20	96-063882	Vehicle	1700 16th St S
X21	96-063883	Vehicle	1800 16th St S
X22	96-063063	Beauty Supply Warehouse	2320 9th St S
X23	96-063884	Dwelling	1523 9th St S
X24	96-062885	Dwelling	1509 9th St S
X25	96-063012	Accurate Locksmith (CWA case)	1326 9th St S
X26	96-063878	Occupied Dwelling (Attempt O#96630)	727 26nd Ave S
X27	96-063097	Vehicle	2402 2nd Ave S
X28	96-063879	Three Brothers Grocery (Attempt)	1040 16th St S
X29	96-063880	Dwelling	1101 12th St S
X30	96-063326	Dwelling	2321 3rd Ave S
X31	96-063230	Maximo Shopping Center (Attempt)	4275 34th St S
X32	96-063221	Vacant commerical Building	3100 1st Ave S
X33	96-063326	Dwelling (1/2 Add)	310 23rd St S
X34	96-063326	Dwelling	2325 3rd Ave S
X35	96-062944	Twin Brooks Golf Course	3800 26 Ave S
X36	96-063024	Vacant commerical building	1800 9th St S
X37	96-063483	Accurate Locksmith (Attempt)	1326 9th St S
X38	96-063749	Jordan Park Human Resource Center	1201 22nd St S
X39	96-063741	Jordan Park Resource Center	1124 22nd St S
X40	96-063437	Dwelling	4725 3rd Ave S

Map 4. This map depicts south St. Petersburg, Florida. The offset yellow squares mark arson locations occurring November 13-17, 1996. The "X" number of each icon relates to the browser file at the left and does not reflect the order of occurrences.[3]

NUM	Offnmbr	Name	Address
X41	96-063836	Vehicle	1101 22nd Ln S
X42	96-062937	Tony's Meat Mkt ATT	1209 4th St S
X43	95-063260	Phillip Ent. ACCIDENT	1619 18th Ave S
X44	96-063260	Vact Dwell ACCIDENT	1746 Prescott St S

---

[3] Map and incident information courtesy of Detective R. S. Moland, C.I.S. Investigative Analysis, St. Petersburg Police department.

SUMMARY AND LESSONS LEARNED

The initial civil unrest of October 24 caught most St. Petersburg residents and public safety agencies off guard. Property damage from both the October 24 and the November 14 disturbances was estimated at over $6 million. Much of the damage resulted from the approximately 66 arson fires involving commercial and residential structures and vehicles. The emergency plans that had been used on several occasions for weather emergencies transferred successfully to the civil disturbances. These established plans and the ongoing networking efforts among fire and police within the city, county, and State allowed for the successful mitigation of each occurrence. Also, the component parts of a successful outcome for any emergency--planning, leadership, and organization--were in place.

In reviewing the public safety response to the two incidents, several findings were made:

- Both the Police and Fire and Rescue Departments performed consistent with their plans

- St. Petersburg Fire and Rescue benefited from the flexibility of the Incident Command System (ICS) as they addressed the civil disturbances.

- The fire chief provided the operations officer sufficient support--staffing, apparatus, equipment, and police escort--to successfully implement a tactical plan to handle the Fire and EMS requirements.

- Fire department personnel were sufficiently trained in the areas of firefighting and emergency medical response. However, training had not prepared fire personnel for the psychological impact of working in a hostile and violent environment. The level of violence upset many responders.

The incidents caused only few injuries to civilians, police officers, or firefighters. The discipline of police officers in adhering to policies relative to returning gunfire and the effort of firefighters to maintain constant response for fire and medical emergencies were noteworthy.

1. **The use of a comprehensive Incident Command System is vital to mitigating civil disturbances.** Because of the quickly escalating nature of the civil disturbance, the need to adjust normal response patterns, and the changes necessary in fire suppression tactics, the ICS becomes essential to enable an organization to manage multiple priorities and use existing mutual aid plans to their maximum capability.

2. **Briefings for companies before they engaged in operations contributed to the safety of all emergency responders.** St. Petersburg Fire and Rescue prepared personnel for operations within the hostile areas by briefing all members prior to entry. Specifics regarding operations, tactics, communications, and firefighter safety were discussed. Following a task force's return to the staging area, members were also debriefed regarding their assignment and fitness for duty.

3. **Pre-incident plans were key to the quick response to handle the civil disturbance fires.** The existence of clearly defined emergency plans allowed police and fire to fulfill their respective roles and responsibilities, allowing coordinated interaction between the agencies. These plans included procedures for calling in additional personnel and resources to reinforce existing resources. The St. Petersburg Fire & Rescue policy embraced the NFPA 1500 Guidelines for Firefighter Safety During Civil Disturbances.

4. **Fire departments may want to consider purchasing personal protective gear for emergency responders, and body armor should be mandated as basic equipment whenever response is to an incident that might be or become violent.** St. Petersburg did not have sufficient body armor to issue vests to all personnel operating within the disturbance area. EMS personnel were given preference during the issuing of available gear during the incidents.

There are a few points to consider when contemplating the acquisition of personal protective gear. First, it is important to keep in mind the difference between (1) soft body armor, (2) flack protection, and (3) blunt trauma protection. When addressing the issue of vest protection, a risk assessment for the specific threat should be performed.

a. **Soft Body Armor:** This type of vest is designed to offer ballistic protection under specific circumstances against specific threats. Body armor is not bulletproof and cannot make the wearer immune to all potential threats. Soft body armor will not stop rifle projectiles, special purpose or armor-piercing projectiles, and sharp-pointed instruments such as arrows, ice picks, and knives.[4]

b. **Flack vest:** This type of protection has been used to protect U.S. service personnel since World War II and the war in Vietnam, and it is still used to protect present day service personnel. Flack vests are designed to protect personnel from flack resulting from exploding bombs and pieces of metal flying at high velocity. This type of vest is not designated to protect responders from bullets.

c. **Blunt Trauma. Protection:** This type protects the wearer from close-range blunt trauma injuries such as baseball bats, bricks, and so forth.

Each model of personal protection is designed to stop specified projectiles at specified speeds under guidelines established by the National Institute of Justice (NIJ).

If a department plans to order vests, it must be remembered that women's vests are designed to the meet the comfort and protection needs of a woman and need to be fitted for the needs of female first responders.

5. **Fire department activities were conducted separately from police activities.** The commanding officers of the St. Petersburg Fire and Rescue had given considerable thought to keeping the Fire Department separate from the Police. Fire stations were not used for any type of police activities, surveillance, staging, or command posts. Existing Fire Department procedures for civil unrest also carefully separated Police' and Fire services.

6. **Cultural sensitivity training should be provided to all emergency responders.** The St. Petersburg Fire and Rescue has recognized the need for increased levels of training in the area of cultural sensitivity. SPER plans to provide training to all personnel to increase members' awareness and recognition of special needs in different communities.

---

[4]There is no such thing as a bulletproof vest. Any vest hit enough times or by a high-velocity penetrating round can fail. To help determine how much stopping power is needed against various weapons and bullets, the National Institute of Justice (NIJ) developed a Threat Level Matrix. NIJ certification tests are performed at approved independent labs. During certification, vests are shot both dry and wet and at various agencies. Each vest is placed against a soft clay surface and shot in a predetermined pattern to test for back face deformation that can cause blunt trauma. (Since a bullet hits with a sledgehammer impact, blunt trauma can debilitate one's ability to function.)

7. **A positive fire department image in the community can help prevent harm to firefighters during civil disturbances.** The events in St. Petersburg suggest that firefighters and emergency medical technicians may have avoided becoming targets in part because of their popular image, even among the minority community. The separation of police and fire activities allowed fire crews to work quickly with less fear of attack. No fire or EMS unit reported being fired upon, although as discussed previously in this report, the first engine to arrive at the October disturbance--Engine 3--was attacked.

8. **The media was kept well informed with frequent updates, thus ensuring that information about the incident was based on fact and not on speculation or rumor.** Local officials routinely updated the media via press conferences that were scheduled four times a day. Available at each briefing were the mayor, fire chief, and police chief. An effort was made to explain that Fire and Rescue personnel could not enter areas that had not been secured by the police. SPER dealt quite successfully with issues regarding their response.

9. **Following any major incident, SPFR managers need to monitor all personnel closely for symptoms of post-traumatic stress and encourage them to take advantage of resources available through their employee assistance program or similar health programs.** The United States Fire Administration's 1994 (FA 142/February 1994) report regarding post incident needs states that the initial priorities should be to restore stability and provide aid to those injured and suffering loss caused by the disturbance. Fire service organizations should remember when addressing post-incident needs to include the emotional and psychological needs of its members. Departments should consider establishing stress management programs that address specific needs, including

   • Training of members to recognize stress symptoms and stress reduction techniques;

   • Critical incident debriefing teams; and

   • Short-term counseling and peer support programs.

10. **The fire service may not fully recognize the benefits they derive from their community programs.** Fire prevention, public education, and community outreach efforts go a long way toward maintaining the positive image enjoyed by most fire and rescue departments, and could save firefighters' lives during civil disturbances.

Subsequent to the second incident, the police and fire departments returned to their community-based programs throughout the city, attempting to reach out to the neighborhoods with a greater understanding of deep-rooted social issues of concern to some of the minority citizens. Sensitivity training was added to the SPFR training for new recruits, including awareness training for potential civil, unrest.

Finally, fire and police departments may be well advised to revisit existing community programs to address any special needs in certain neighborhoods. The adoption of community awareness training programs, like those that have been used by police departments elsewhere, may assist with ongoing efforts to improve relations with civilians.

continued on next page

Successful disturbance control depends upon quick and decisive actions built upon plans and preparation. All pre-incident planning for fire response should prepare for the worst-case scenario in an effort to maximize personnel safety. Afterward, healing must begin, both within the community, and within the public safety agencies that responded. Reestablishing relationships within the community is clearly a priority. Fire department officials can face animosity, mistrust, residual tensions, and frustration from emergency personnel whose safety was placed on the line during the outbreaks of violence. Sensitive past incident counseling sessions may help resolve some of these issues.

APPENDICES

Appendix A: Command Structure, October 24-25, 1996

Appendix B: Command Structure, November 13-14, 1996

Appendix C: Incident Review: Dispatch Procedures

Appendix D: Fire Department 600 Series,
Standard Operating Procedure No. 600-24
Subject: Civil Disturbance

Appendix A

COMMAND STRUCTURE CIVIL UNREST

THURSDAY, OCTOBER 24, 1996

Command established: 2108 10/24/96
Command terminated: 0345 10/25/96
Call processed through command: 96

Command: Assistant Chief, St. Petersburg Fire and Rescue
Fire operations: Rescue 5 / Rescue Lieutenant / District Chief 27 (Indian Rocks FD)
Fire operations: Rescue 5 / Rescue Lieutenant / District Chief 29 (Seminole FD)
EMS operations: Rescue Lieutenant / Captain
Staging: Command
Logistics: Assistant Chief with help from city
EOC communications: None
Call screener: Division Chief / Rescue Lieutenant
Public information officer: Division Chief
Documentation: None
Rehab: Rehab 44 (Belle Air FD) / Rehab 49 (Clearwater FD)
Water supply: None
Air recon: Captain
Air supply: Shop Manager

Task force #1
District Chief #5, District Chief
Engine #5
Engine #1
Truck #1

Task force #2
District Chief #10, Captain
Engine #3
Engine #4
Truck #11

Task force 3
Division Chief
Captain
Engine #11
Engine #8
Engine #18 (Lealman FD)

Task force #4
District Chief #18 (Lealman FD)
Engine #10
Engine #6
Rescue #10

Task force #5A
District Chief #35 (Pinellas Park)
Engine #9
Truck #23 (St. Pete Beach FD)
Engine #36 (Pinellas Park FD)

Task force #5B
Division Chief
Pumper #12
Truck #1
Engine #7

Task force #6
Engine #13
Truck #9
Truck #19 (Lealman FD)
Pumper #12

Rescue (ALS) teams
Rescue #11
Rescue #8
Rescue #5

Moveups to provide coverage
Engine #10 to Master Fire Station
Rescue #10 to Master Fire Station
Engine #18 (Lealman FD) to Station 3 with a divert to Master Station
Rescue #7 to Station 4
Engine #33 (Pinellas Park FD) to Station 10
Engine #35 (Pinellas Park FD) to Station 4
Engine #38 (Largo FD) to Station 33 (Pinellas Park FD) with divert to Station 7
Engine #40 (Largo FD) to Station 36 (Pinellas Park FD)
Engine #7 to Master Station

Appendix B

COMMAND STRUCTURE CIVIL UNREST

WEDNESDAY, NOVEMBER 13, 1996

Command established: 1848 11/13/96
Command terminated: 0657 11/14/96
Calls processed through command: 146

Command: Assistant Chief
Fire operations: Lieutenant
Fire operations: Division Chief
EMS operations: Lieutenant / Captain
Staging: Captain / Lieutenant
Logistics: Lieutenant
Communications: Communications #38 (Largo FD) and personnel from County 9-1-1
Call screener: Lieutenant
Public information officer: Assistant Chief
Documentation: Assistant Chief (St. Pete Beach FD)
Rehab: REHAB 44 (Belle Air FD)
Water supply: None
Air recon: None
Air supply: Squad #32 (Seminole)

Task force #1
District Chief 45
Engine #1
Engine #5
Truck #1

Task Force #2
District Chief
Engine #7
Engine #11
Truck #11

Task force #3
District Chief #18 (Lealman FD)
Engine #3
Engine #1 8 (Lealman FD)
Truck #9

Task force #4
Assistant Chief (Lealman FD)
Engine #8
Engine #40 (Largo FD)
Engine #10
Truck #19 (Lealman FD)
Engine #13

Task force #5
District Chief #35 (Pinellas Park FD)
Engine #40 (Largo)
Truck #23 (St. Pete Beach FD)
Engine #6
Engine #8

Task force #6
District Chief #41 (Largo FD)
Engine #49 (Clearwater FD)
Engine #13
Truck #41 (Largo FD)

Task force #7
Assistant Chief (Seminole FD)
Engine #28 (Indian Rocks FD)
Engine #4
Truck #20 (South Pasadena FD)

Task force #8
District Chief #48 (Clearwater FD)
Squad #32 (Seminole FD)
Pumper #12
Engine #35 (Pinellas Park FD)

Task force #9
Battalion Chief #3 (Tampa FD)
Division Chief (Clearwater FD)
Engine #9 (Tampa FD)
Engine #10 (Tampa FD)
Ambulance #9 (Tampa FD)
Rescue #14 (Tampa FD)

Rescue (ALS) teams
Rescue #1
Rescue #3
Rescue #4
Rescue #10

Move up to provide coverage
Engine #32 (Seminole FD) to Station 9
Rescue #33 (Pinellas Park FD) to Station 4
Engine #38 (Largo FD) to Station #6
Engine #39 (Largo FD) to Station #13
Engine #58 (East Lake FD) to Station #18 (Lealman FD)
Engine #68 (Palm Harbor FD) to Station #10
Engine #62 (Dunedin FD) to Station #33 (Pinellas Park FD)
Truck #29 (Seminole FD) to St. Petersburg Master Station

Appendix C

St. Petersburg Civil Disturbance
Incident Review
Dispatch Procedures

(Written by George Buck)

TABLE OF CONTENTS

Purpose

This report is intended to provide the reader with a general overview of SPFR dispatch functions during St. Petersburg's civil disturbance of October 24/25 and November 13/ 14, 1996. Specific details peculiar to each day of unrest will be addressed from a global perspective as they relate to identified strengths and weaknesses.

For purposes of this report, references made to Task Force units include Rescue Extraction Teams that operated under the Rescue Operations leg of Dome Command. Dispatch procedures for both fire and rescue operations were similar in many respects.

Six critical questions establish the format and make up the content of this report. This writer's personal observations, recommendations, and conclusions are included and do not necessarily represent the opinions of the management of St. Petersburg Fire & Rescue.

Overview of Dispatch Functions

Dispatch operations changed to some degree each day of the civil disturbance. However, there were some common factors that contributed to a successful operation.

- Emergency responses to the area of disturbance (i.e., No-Fly-Zone) were altered by Command staff to provide a measure of safety for Fire/Rescue personnel.

- All 9-1-1 calls to the no fly zone were redirected to the Incident Commander via digital pager for processing and dispatch. Task Force units were dispatched by command staff assigned to tactical channels and in concert with law enforcement agencies.

- A law enforcement liaison was assigned to the command post to provide current situation reports on disturbed areas to facilitate adjustments to the perimeter of the no fly zone.

- CCD attempted to monitor tactical channels to document responses and times.

- Efforts were made to establish a communication link with CAD to monitor status screens and assist command staff with move-up recommendations.

An air reconnaissance unit via the Sheriffs Office was utilized to provide the Incident Commander and fire units with tactical information on reported structure fires, safe travel routes in and out of disturbed areas, and confirmation that the area was secured by law enforcement. However, this operation was discontinued after Day 1 because the helicopter took gun fire and sustained damage.

Dome Command Statistics

	Oct. 24, 1996	Oct. 25, 1996	Nov. 13, 1996	Nov. 14, 1996
Command Established	2108 10/24/96	1733 10/25/96	1848 11/13/96	1840 11/14/96
Command Terminated	0345 10/25/96	0200 10/26/96	0657 11/14/96	0037 11/15/96
9-1-1 Calls Processed by Command	96	16	146	22

How Did CCD Respond To Our Request To Establish A 'No-Fly-Zone'?

It was generally understood by CCD that the 'Hot-Zone' (later termed the No-Fly-Zone) represented a disturbed area unsecured by law enforcement. Although the term 'no-fly-zone' isn't typically associated with Fire/Rescue activity, it seemed to be appropriately and consistently applied by all concerned.

After establishing the no fly zone, all of the 9-1-1 calls for help within the affected area had to be managed differently by CCD because of the following:

- Automatic dispatching of available units into the no fly zone was prohibited.

- Adjustments were made to predetermined responses to provide a greater level of safety to all personnel.

- Ambulances were restricted from the area.

- Unusual high volume of active calls slows down the CAD.

- Accidental dispatch of Fire/Rescue unit into affected area could jeopardize the safety of personnel.

Consequently, it became extremely important to identify any calls within the no fly zone. To accomplish this, CCD added specific caution notes to each grid within the affected area that would alert the dispatcher to verify the location prior to dispatching.

Initially this took a little longer than usual because the 'multi-grid caution note' capability of the CAD was out-of-service. As a result, CCD had to add caution notes manually to approximately 12 to 15 grids. Prior to completing this task, the call volume for the no fly zone dramatically increased, requiring the dispatchers to cross-reference addresses manually with current boundaries and override the CAD by attaching a single unit to the call (SP-200 or SP-LR1).

Once caution notes were entered into the no-fly-zone grids, it became easier to identify calls within the affected area. Yet, this process was not completely automatic because several grids on the perimeter extended outside of the no-fly-zone boundary. Thus, calls on the perimeter of the no fly zone still needed to be cross-referenced to determine whether to dispatch the call normally or to special up SP-200 or SP-LR1.

Every effort was made to keep CCD informed of current boundaries for the no fly zone.

All 9-1-1 calls received by CCD for locations within the no fly zone were forwarded to SPFR Dome Command for processing. To verify the call was sent via digital pager and that the page bridge was operative, dispatchers were instructed to look for the time stamp in the notes of each call prior to closing the call.

Critical Observations:

- Manually identifying 9-1-1 calls within the no fly zone accurately was contingent upon dispatchers ability to cross-reference the address or their familiarity with the location. Even after caution notes were added to the affected grids, calls to the perimeter of the no fly zone required dispatchers to manually verify the location against current no-fly-zone boundaries.

- It was critically important to keep CCD informed of current boundaries for the no fly zone. These notifications play an important role in establishing a safe perimeter for Fire/Rescue personnel working outside of the no fly zone.

- Caution notes attached to specific grids require the dispatcher to pull up the notes manually.

Recommendations:

- CCD has a staff of experienced dispatchers familiar with the geography and layout of St. Petersburg. There were no incidents of inappropriate dispatches into the no fly zone. Yet, considering the potential and the possible ramifications of a dispatch into the no fly zone, I would recommend that a member of SPFR command staff be available to respond to the Communication Center (as outlined in our Civil Disturbance Plan) and offer assistance as needed.

- Keep CCD informed of current boundaries for no fly zone with timely updates.

- Develop an automatic pop-up screen for caution notes that require the dispatchers to override manually rather than manually retrieve.

Conclusions:

- CCD performed admirably, given the unusual circumstances and demands placed upon them.

- Implementation of the no fly zone was handled smoothly and efficiently.

- Redundant systems were put in place to ensure that 9-1-1 calls were transferred to Dome Command.

- No-fly-zone grids were adjusted as necessary to reflect current boundaries without too much difficulty.

What Problems Did CCD Encounter With The Development Of Task Forces?

The Task Force concept was originally comprised of two engines, one truck and one command staff officer--otherwise known as the Task Force Leader. This configuration changed as the availability of truck companies dwindled.

To facilitate this request, CCD encoded units requested by Dome Command and manually assigned them to each Task Force developed. This process was not complicated or cumbersome. It simply required specializing the requested unit to the appropriate call and tactical channel.

As events unfolded on the first day of unrest, Task Forces were developed in pace with the demand. During the first few hours, it was extremely difficult to predict the magnitude of the incident. Thus, Task Force units were put together as the need arose. However, after it became clear the incident would be quite extensive, requiring the assistance of outside agencies, an attempt was made to develop Task Forces in advance and position them at the Dome staging location.

It was important to request units that were not already committed to other assignments to fill Task Force needs. Access to the CAD in the Command Bus aided staff in making decisions that would ultimately affect resources available to the entire county. Additionally, having personnel familiar with the CAD (CO-701) in the Command Bus assisted with the development of Task Forces and served to free Fire/Rescue staff to address command functions rather than being concerned with unit availability and move-ups.

However, one of the problems encountered with the development of Task Forces was the need to backfill districts and stations with outside agencies. At one point, the call for Task Forces came so quickly that it overran the ability of CCD to move units automatically to cover. To address this, CO-701 in concert with Dome Command staff manually moved units to provide adequate coverage for the city.

Critical Observations:

- Manually specializing units to comprise Task Forces was to some degree dependent on Dome Command's knowledge of the units that were available in the county at any given time. Requesting a unit that was already committed or otherwise out of service only served to complicate and slow down the process.

- Access to the CAD in the Command Bus was a valuable tool for developing Task Forces and monitoring resources available to the remainder of the city and county.

- The on-site assistance of CCD personnel (CO-701) provided command staff with the technical assistance needed to develop Task Forces on the fly and maintain an adequate level of coverage for the rest of the city.

- An effort was made to have a SPFR unit with each Task Force developed. As the demand increased, it became necessary to reassign some units to different Task Forces. This caused some confusion for CCD who attempted to monitor unit status and Tactical channels.

Recommendations:

- Explore the possibility of developing alternative response run cards that assign Task Force and/ or Strike Team units in advance. Build cards of sufficient quantity and depth to provide initial resources automatically.

- Explore the communication technology that is available to create a reliable link between the Command Bus and CCD (e.g., Cellular modem, line-of-sight infrared, and radio technology).

- Build into SPFR Civil Disturbance Plan the request for a CCD liaison to respond to the Command Bus or SPFR command staff officer report to CCD.

- Make efforts to keep CCD informed of unit status (i.e., Task Force assignment, availability, staging, rehab, etc.). This can serve as a redundant system for personnel accountability and safety.

Conclusions:

- Transition to an altered response (Task Force) was accomplished without incident. CCD responded to our request quickly and efficiently.

- The assistance of CO-701 was valuable.

What Difficulties Did CCD Have Monitoring Tactical Channels?

For the most part, our primary tactical channels were 1-G, 1-I, and 1-J. Dome Command operated from 1-H, as did staging during the initial phase. This really didn't present any operational concerns for CCD. Aside from having to redirect 9-1-1 calls outside of the no fly zone that were automatically dumped to 1-G, business carried on as usual. In fact, activity for the rest of the city and county was relatively quiet during the civil disturbance in St. Petersburg.

CCD did experience some difficulty monitoring and recording the status of Task Force units operating on tactical channels. During the initial phase of the operation, many of the 9-1-1 calls were closed after being sent to Dome Command for processing so as not to slow down the CAD. At the discretion of Dome Command, these calls were held until law enforcement agencies could confirm that the area was secure.

However, once the area was deemed secure, Dome Command would dispatch a Task Force to the location and assign those units to a tactical channel. Task Force units essentially responded to incidents that from CCD's perspective were closed and no longer existed. Unless Dome Command informed CCD to recreate the call, there was no way for CCD to track the units that responded, their status, and response times. This was true even though CCD monitored the tactical channel.

That situation was corrected during the subsequent days of unrest. 9-1-1 calls were created, forwarded to Dome Command, and then closed after a short period time. Once the area was deemed secure, Dome Command would ask CCD to recreate the call and assign the designated Task Force. Unit status and response times were captured.

Every effort was made by CCD to monitor tactical channels as normal. Unfortunately, this created a little confusion because command staff personnel unconsciously took on the role of dispatch. All the 9-1-1 calls forwarded to Dome Command were eventually dispatched by command. Task Force units were hailed on 1-H by Dome Command and instructed to switch to the appropriate tactical channel for assignment. Each tactical channel was overseen by command staff personnel who received the particulars of the call verbally from Dome Command, dispatched the call, and relayed pertinent information to Task Force units.

Consequently, when Task Force units would hail dispatch, command staff personnel who actually dispatched the call would respond in addition to CCD. Understandably, a measure of frustration was felt by some of the CCD dispatchers who were not sure of the role they were to play as they monitored tactical channels.

An additional problem encountered by CCD involved the Simms Code H system. This system was essentially rendered ineffective because

- A large quantity of spare radios was utilized. It would have been too time-consuming and confusing to adjust identifiers to match assigned units.

- There were a number of units operating in the no fly zone from agencies outside of Pinellas County without Simms Code H capability.

- CCD experienced periods of 'Fail-Soft' on several of Dome Command's operating tactical channels.

- CCD experienced difficulty keeping track of which units were assigned to which Task Force.

From an operational perspective, use of the Simms Code H system during civil unrest does not adequately identify units or personnel in distress, and it renders the radio open (i.e., operator cannot transmit or receive) until reset. As a result, at 2112 hours on the 1st day of unrest, Dome Command replaced the use of the Simms Code H by advising all personnel to transmit their last name followed by the term 'Mayday.'

Critical Observations:

- A measure of confusion and frustration developed as CCD attempted to fulfill their responsibility as dispatchers, only to find that their role had been inadvertently supplanted by command staff.

- CCD experienced difficulty monitoring and recording the status of Task Force units operating on tactical channels.

- The Simms Code H system did not adequately serve Fire/Rescue personnel during a civil disturbance.

- 1-G was the primary tactical channel used during the incident. This channel is also the primary tactical channel for normal Fire/Rescue operations in St. Petersburg. Calls outside of the no fly zone for SPFR would automatically default to 1-G, thus requiring CCD to manually redirect.

- Not all Fire/Rescue units within the county have the same low priority channels available. While fleet 2 was available for use, it too was not uniformly programmed into all Fire/Rescue units within the county.

- Dome Command's operating tactical channels experienced 'Fail-Soft' on several occasions, interfering with Command's ability to coordinate resources.

Recommendations:

- Provide clear directions to command staff assigned to monitor tactical channels. Such personnel should function more in line with the ICS description of an operation, division, or branch officer rather than dispatch.

- Notify CCD when Task Force units are assigned to a closed call. Have the call recreated and monitored as normal by CCD. Additionally a Command staff officer should be assigned to monitor each tactical channel to address any strategic and tactical concerns.

- Establish a uniform countywide policy to address the use of tie term 'Mayday' in conditions not conducive to Simms Code H activation.

- Program three low-priority channels common to all Fire/Rescue units within the county capable of being monitored by CCD for disaster and civil disturbance use.

- Explore circumstances that caused 'Fail-Soft' on several operating tactical channels and seek remedy.

Conclusions:

- Command staff assigned to monitor tactical channels (this writer included) hampered CCD ability to capture critical information for documentation purposes.

- CCD stood ready to assist on tactical channels as normal. Unfortunately, SPFR chose to assume dispatch functions beyond what was necessary. It should be noted that this situation could have easily tipped to the other direction. If emergency activity in the rest of the city and county were above normal, SPFR quite likely could have been forced to handle this incident as a Condition 5 exercise (i.e., assuming all dispatch functions).

What Could Have Been Done To Identify Duplicate Calls More Effectively?

One of the challenges that CCD faced during the civil disturbance was identifying duplicate calls prior to forwarding them to Dome Command. To this date it is still undetermined how many of the 9-1-1 calls processed through command were duplicates. Yet, every effort was made to filter out duplicate calls.

The CAD is programmed to identify possible duplicate calls by comparing the incoming 9-1-1 call's location with all active 9-1-1 calls. A prompt on the call-taker/dispatcher's screen will appear that indicates a possible duplicate call. This prompt also includes a code for the active call so that the call-taker can easily cross-reference it with the incoming call by nature and location.

This system of checks and balances works only when calls are active. During the civil disturbance, 9-1-1 calls were closed relatively quickly because

- The list of active calls steadily increased to the point where dispatchers had to scroll through several screens to find a call.

- The large number of active calls slowed the CAD down to an unacceptable speed.

- 9-1-1 calls sat idle without assigned units until Dome Command was confident the area was secured by law enforcement.

Thus, with the call already closed and forwarded to command, the CAD was unable to identify duplicate calls.

Every effort was made at CCD to identify duplicate calls manually. Call-takers and dispatchers would communicate among themselves as calls came in to determine if any had received a call of a similar nature in the same location. Of the dispatchers interviewed, several indicated they were able to identify a significant number of duplicate calls.

Similar efforts to identify duplicate calls took place in the command bus. Although command staff felt somewhat impelled to send response units to each call received, there were some calls that were obvious duplicates. These calls were usually handled by the units arriving to the first address and confirming secondary addresses as duplicates.

Critical Observations:

- CCD had difficulty identifying duplicate calls after calls were forwarded to Dome Command and closed.

- It was very difficult for Command staff to identify duplicate calls.

- Command staff felt a little reluctant to categorize calls as duplicate once an incident number was assigned.

Recommendations:

- Create a running list of 9-1-1 calls by location and nature to be cross-referenced by command staff to help identify duplicate calls during high call volume.

- Explore the possibility of establishing a minimum time frame that calls will remain active prior to closing out to help identify duplicate calls.

Conclusions:

- It is extremely difficult to evaluate how effective CCD was in identifying duplicate calls because there is no way to determine how many CCD caught. However, based on the energy and effort put forth by CCD to help reduce the number of unnecessary calls, they did an excellent job. Without the assistance of the CAD, call takers and dispatchers manually cross-referenced addresses and call nature codes to identify duplicate calls. Much of this was done from memory.

- On several occasions Dome Command was inundated with 9-1-1 calls they were forced to hold. It is difficult to imagine how busy command staff would have been if CCD had not made any effort to filter out duplicate calls.

How Efficient Was Sending 9-1-1 Calls to Command Via Digital Pager?

After CCD identified a 9-1-1 call requiring response into the no fly zone, it was forwarded to Dome Command for processing. These calls were sent to SP-200 and SP-LR1 in the command bus via digital pager. Calls were then recorded in chronological order and held until law enforcement could verify the location was secure.

These digital pagers reached their maximum memory very quickly as the pace of 9-1-1 calls into the no fly zone increased. As a result, a handwritten chronological list of 9-1-1 calls was the only permanent record available to Dome Command by which Task Force units could be assigned.

After some time and considerable effort, a cellular link with CCD was established which provided Dome Command with 'hard copy' of each 9-1-1 call prior to its being closed. Communications 38 provided this valuable link. However, the technology available did not allow CCD to ship the calls to that unit. The cellular link enabled access to the status screen and each call was brought up manually and printed, using the print screen option of the PC. The 'hard copy' produced was extremely helpful. Yet, because the print screen option was so time consuming, more than a few 9-1-1 calls were not printed before CCD closed them.

On Day 2 and thereafter, a cellular link with CCD was established in the command bus via laptop computer. This gave command staff access to the status screen (no print option) and created a measure of redundancy, which was needed to verify receipt of 9-1-1 calls.

Critical Observations:

- 9-1-1 calls forwarded to either SP-200 or SP-LRI digital pager during the civil disturbance required the personnel either to (1) commit to the Command Bus for the duration, (2) assign themselves a spare pager, or (3) go without a digital pager. Command staff personnel can be called upon to fulfill several assignments during the course of an incident. The ability to communicate with such personnel is essential.

- Receipt of 9-1-1 calls was for a period of time completely dependent on the CCD's page bridge. This system has gone down in the past on several occasions without warning. The command bus needs a redundant communication system installed that will not affect 800 MHz radio traffic. Fire stations have redundant communication systems built: 800 MHz radio, station encode and printout, and digital pager. The command bus has 800 MHz radios only.

- Radio traffic would have escalated to an unmanageable level had the page bridge failed, leaving CCD unable to digitally forward 9-1-1 calls.

- The Command Bus was not properly equipped to handle an incident of this magnitude.

Recommendations:

- Make available to command staff several spare digital pagers that can be used to receive multiple or consecutive messages, or both.

- Explore the possibility of wiring a telephone (land-line) quick connect in the Command Bus that can be easily attached by the telephone company to any telephone service pole.

- Explore the possibility of purchasing necessary computer equipment to maintain a reliable communication link with CCD.

- Explore existing cellular technology to determine if an encoder system (similar to fire station encode system) could be installed in the command bus to receive dispatch notification and printouts.

- Explore the possibility of purchasing a fax machine as a redundant communication link between CCD and the command bus.

- Explore possible funding options with the Pinellas County to equip the command bus properly for large-scale operations within the county.

Conclusions:

- The command bus was ill equipped to handle incidents for which it was intended to be used.

- Command staff managed a large volume of 9-1-1 calls very efficiently despite the lack of modern computer, cellular, and radio technology.

How Reliable Was Cellular Technology In Maintaining A Link With The CAD?

The need to establish communication link with the CAD was evident after the first day of operation. Command staff experienced some difficulty keeping track of in-coming 9-1-1 calls from CCD and assembling Task Force units without being able to access the status screens.

On Day 2, communication 38 was called to the command post location to provide this needed link with the CAD. However, computer hardware, software, and cellular problems hampered this effort. CO-711 responded to the Command Bus and provided staff with a laptop computer in hopes of creating a cellular link with the CAD. Once again, this was delayed because of hardware conflicts.

Eventually, communication 38 established a link with the CAD and provided Dome Command with a hard copy of each 9-1-1 call in the no fly zone as it was forwarded by CCD. It was not possible for command staff to monitor the CAD status screens in communication 38 because of the unit's small size.

CO-701 provided a laptop, cellular phone, and appropriate hardware to maintain a link with the CAD in the command bus. This link was interrupted only on a few occasions when batteries became weak and needed to be exchanged. Aside from these few momentary difficulties, the cellular link provided by CO-701 was reliable and helpful to staff operating in the command bus.

Critical Observations:

- The use of hand-held battery-powered cellular phones was problematic at best during Day 2 and thereafter. The cellular phones used did not have the capability of being online and connected to a charger at the same time. Thus, after several hours of operation, the batteries became weak and we eventually lost our connection.

- The cellular link to the CAD was affected by signal strength. It is undetermined if the command bus location had any effect on the cellular signal strength. On several occasions, the cellular link with the CAD was broken as a result of a loss in signal strength. Efforts were made to boost signal strength by relocating the cellular phone antennae near a window or in contact with an outside metal conductor, with no results.

- The cellular phones permanently installed in the Command Bus could not be used to establish a link with the CAD. The hardware would not allow connection to a PC.

- The EOC on several occasions requested statistical information from Dome Command. This information was relayed by word of mouth and were thus was subject to error. A fax machine would provide a method to transmit documents that would essentially eliminate this potential for error. Additionally, a fax machine would provide another method of communicating with the Command Bus.

Recommendations:

- Explore the possibility of purchasing three cellular fax/modem devices to be hardwired in the Command Bus. These devices should be easily connected to a PC for fax and data communication.

- Explore the possibility of purchasing portable computer equipment to be used in the Command Bus and elsewhere during down time.

- Purchase a fax machine for the Command Bus.

Conclusions:

- The communication link with the CAD was a valuable tool for command staff. After identifying the problems and working out the bugs, command staff was able to maintain a relatively reliable link with the CAD. Unfortunately, this was toward the end of the incidents.

- The Command Bus needs to be equipped to receive 9-1-1 calls from CCD. During Condition 5 exercises, the Command Bus should be capable of functioning as a mobile communication sub-center.

Appendix D

FIRE DEPARTMENT 600 SERIES

STANDARD OPERATING PROCEDURE

No. 600-24

SUBJECT: CIVIL DISTURBANCE

Upon initiation of any moderate to large-scale civil disturbance, the following guidelines are to be considered in the decisionmaking process when responding to fire and rescue-related emergencies during such disturbances. These guidelines are divided into the following areas:

Philosophy, Organization, Communications, Operations, and Recovery.

1. **PHILOSOPHY**

   1.1. Fire and rescue department forces are not to be used for disturbance control or combative intervention against the perpetrators.

   1.2. At all times, decisions are to be made in the interest of reasonable degrees of safety for responding personnel and equipment, balanced against the department's responsibility to the public for emergency fire and EMS response.

2. **ORGANIZATION**

   2.1. The management system used to mitigate the incidents will be in concert with S.O.P. 600-1.

   2.2. In the event of moderate buildup of disturbance activities, the activation of SWAT medics should be considered or placed on standby to assist law enforcement so as not to deplete on-duty emergency service personnel capabilities.

3. **COMMUNICATIONS**

   3.1. Consideration will be given to assigning one staff officer from the jurisdiction involved to the County 9-1-1 Communications Center to assist in coordinating the district's resources dispatched from that center, including coordination of apparatus, mutual-aid support, move-ups, and call duplication prioritization.

   3.2. Communications are to be established via the command post with the law enforcement command system by the most practical means, such as radio, telephone, person-to- person, liaison, etc., to ensure efficient and safe operation.

   3.3. Immediately upon recognition of a significant buildup of disturbance activity, district chiefs of the impacted jurisdiction will notify their administrative staff via the County 9-1-1 Communications Center of the situation.

3.4. All operations chiefs and district chiefs will be notified of the activity and their potential involvement via papers by Pinellas County Communications.

3.5. Where practical, and when the perimeter of the impacted area is identified, the remaining areas of the district may be handled in a 'business as usual' dispatch mode. Where possible, identify the impacted areas using existing 9-1-1 grid lines to assist the 9-1-1 Communications Center with dispatching concerns and caution notes.

3.6. Ambulance units assigned to calls within the given higher risk area must report to staging and work through the fire and rescue command structure.

3.7. Ambulance units assigned to calls outside of the given higher risk area shall not knowingly travel through the higher risk area while enroute to the call.

4. **OPERATIONS**

4.1. Response/Staging

4.1.1. Warning lights, sirens, and horns are not to be used normally in responding within impacted area.

4.1.2. The staging of fire and rescue units is anticipated in order to provide fire command with an opportunity to evaluate the safety and security considerations for fire and EMS personnel.

4.1.3. All fire and rescue department mutual-aid units assigned to the incidents, or in staging, are to be fully informed of situations as they develop. The safety of mutual-aid units will receive the same priority as the jurisdiction's units.

4.1.4. Requests to use fire stations for a law enforcement staging or command post or both, or the use of fire department apparatus, must be approved by the impacted fire chief or his designee.

4.2. Tactical

4.2.1. Where appropriate, a heavy stream application followed by a rapid withdrawal may be used. This should be based on reasonable judgment and evaluation of the relative loss/risk factor which shall be used in determining if a fire should be fought or not.

4.2.2. As attaching supply lines to a hydrant impedes the immediate withdrawal of units the use of hydrants will be with the expressed permission of the Incident Commander or Task Force Leader.

4.2.3. In the event of the use of chemical agent for emergency crowd control, normally all fire and rescue personnel will have been withdrawn. In the event that they have not, and have been exposed to chemical agents, decontamination will be accomplished through coordination with the on-duty hazardous material officers.

4.2.4. If Task Force concepts are used, a Task Force leader will be designated by command.

4.2.5. Where practical, at least one person, or preferably one unit from the jurisdiction involved will be a part of each tack force to provide some local familiarity.

4.3. Equipment

    4.3.1. Where practical, backup equipment may be used to prevent damage to first line fire apparatus. This decision must be based on judgment and the urgency of the situation.

    4.3.2. Exterior tools and devices that could be used as weapons against fire and rescue personnel shall be placed inside the vehicle compartments for safety.

    4.3.3. At all times, full protective equipment shall be worn. Company officers shall make the necessary safety decisions for members of their command.

4.4. EMS

    4.4.1. Emergency medical transportation is to be accomplished by 'load and run procedures in high-risk situations, later transferring the patient to Sunstar, if appropriate, outside of the high-risk area. Note: After removing the patient from high-risk exposure, all appropriate pre-hospital medical operating procedures are to be carried out as is normally done.

5. RECOVERY

5.1. Document all cases and information, in detail, for cost recovery and later investigation.

5.2. Give consideration to assigning a documentation officer to track all incident related activity.